TURKEY AS A U.S. SECURITY PARTNER

F. STEPHEN LARRABEE

Prepared for the United States Air Force

Approved for public release; distribution unlimited

The research described in this report was sponsored by the United States
Air Force under Contract FA7014-06-C-0001. Further information may
be obtained from the Strategic Planning Division, Directorate of Plans,
Hq USAF.

Library of Congress Cataloging-in-Publication Data

Larrabee, F. Stephen.
   Turkey as a U.S. security partner / F. Stephen Larrabee.
     p. cm.
   ISBN 978-0-8330-4302-3 (pbk. : alk. paper)
   1. United States—Foreign relations—Turkey. 2. Turkey—Foreign relations—
United States. 3. National security—United States. 4. National security—Turkey.
5. United States—Foreign relations—Middle East. 6. Turkey—Foreign relations—
Middle East. I. Title. II. Title: Turkey as a United States security partner.

   E183.8.T8L37 2008
   327.730561—dc22

                        2007052706

The RAND Corporation is a nonprofit research organization providing
objective analysis and effective solutions that address the challenges
facing the public and private sectors around the world. RAND's
publications do not necessarily reflect the opinions of its research clients
and sponsors.

RAND® is a registered trademark.

Cover design by Peter Soriano

Published 2008 by the RAND Corporation
1776 Main Street, P.O. Box 2138, Santa Monica, CA 90407-2138
1200 South Hayes Street, Arlington, VA 22202-5050
4570 Fifth Avenue, Suite 600, Pittsburgh, PA 15213-2665
RAND URL: http://www.rand.org/
To order RAND documents or to obtain additional information, contact
Distribution Services: Telephone: (310) 451-7002;
Fax: (310) 451-6915; Email: order@rand.org

Preface

Since the creation of the North Atlantic Treaty Organization in 1949, America's security partnership with Turkey has been a strategic asset that both parties value. Now, however, trends in the greater Middle East, in Turkish security policies, and within Turkish society itself appear to be eroding the commonality of interests that constitutes the foundation of that partnership. Left unchecked, these trends could diminish U.S. influence in Turkey and increase instability in the Middle East. This monograph explores the dynamics of the evolving U.S.–Turkish security relationship and their implications for U.S. foreign and security policies.

The research reported here was sponsored by the Director for Operational Plans and Joint Matters (AF/A5X), Headquarters United States Air Force. The work was conducted within the Strategy and Doctrine Program of RAND Project AIR FORCE as part of a fiscal year 2006 study "Risks and Rewards in U.S. Alliances."

RAND Project AIR FORCE

RAND Project AIR FORCE (PAF), a division of the RAND Corporation, is the U.S. Air Force's federally funded research and development center for studies and analyses. PAF provides the Air Force with independent analyses of policy alternatives affecting the development, employment, combat readiness, and support of current and future aerospace forces. Research is conducted in four programs: Aerospace Force

Development; Manpower, Personnel, and Training; Resource Management; and Strategy and Doctrine.

Additional information about PAF is available on our Web site: http://www.rand.org/paf/

Contents

Summary

In the future, Turkey is likely to be an increasingly less-predictable and more-difficult ally. While Turkey will continue to want good ties with the United States, Turkey is likely to be drawn more heavily into the Middle East by the Kurdish issue, Iran's nuclear ambitions, and the fallout from the crisis in Lebanon. As a result, the tension between Turkey's Western identity and its Middle Eastern orientation is likely to grow. At the same time, the divergences between U.S. and Turkish interests that have manifested themselves over the last decade are likely to increase (see pp. 7–14, 17–19).

Given its growing equities in the Middle East, as well as the current strains in U.S.–Turkish relations, Turkey will be even more reluctant to allow the United States to use its bases in the future, particularly the air base at Incirlik, to undertake combat operations in the Middle East (see p. 29). President Turgut Özal's willingness to allow the United States to fly sorties out of Incirlik during the 1991 Gulf War was the exception, not the rule. Since then, Turkey has increasingly restricted U.S. use of Incirlik for combat missions in the Middle East. Thus, the United States should not count on being able to use Turkish bases, particularly Incirlik, as a staging area for combat operations in the Gulf region and the Middle East (see p. 25).

Moreover, given the importance of the Kurdish issue for Turkish security, Turkey has strong reasons to pursue good ties with Iran and Syria (see pp. 11–14), both of which share Turkey's desire to prevent the emergence of an independent Kurdish state. Turkey's growing energy ties with Iran have reinforced interest in that particular relationship.

Thus, Turkey is unlikely to support U.S. policies aimed at isolating Iran and Syria or overthrowing the regimes in either country (see pp. 11–14). Rather, Ankara is likely to favor policies aimed at engaging Iran and Syria and to encourage the United States to open dialogues with both countries (see pp. 11–14).

Turkey's interest in good relations with Iran and Syria represents a potential point of tension in U.S.–Turkish relations and highlights the need for the United States to consult closely with Ankara to try to ensure that U.S. and Turkish policies do not operate at cross purposes. Like the United States, Turkey does not want to see a nuclear-armed Iran. While it does not perceive an existential threat from a nuclear-armed Iran, Ankara fears that Iran's acquisition of nuclear weapons could destabilize the Gulf region and force Turkey to take defensive countermeasures to safeguard its own security (see pp. 12–13).

However, while Turkish officials are concerned about the long-term security implications of a nuclear-armed Iran, Prime Minister Recep Tayyip Erdogan's government is strongly opposed to a military strike against Tehran, which it believes could further destabilize the region. Thus, the United States could not count on the use of Turkish bases in any military operation against Iran. Indeed, such a strike could provoke a serious crisis in U.S.–Turkish relations and significantly exacerbate current strains with Ankara (see p. 13).

In the near term, however, the most important source of potential discord between the United States and Turkey is likely to be over how to deal with the terrorist attacks the Kurdistan Workers' Party (PKK) conducts from sanctuaries in northern Iraq (see pp. 7–11). The number of Turkish security forces the PKK has killed has risen dramatically over the last year. Domestic pressure, especially from the Turkish military, has been growing for Turkey to take unilateral military action against the PKK. The landslide victory by the Justice and Development Party in the July 22, 2007, parliamentary elections has strengthened Erdogan's hand politically and bought him some breathing room diplomatically. But if the attacks intensify in the aftermath of the elections, Erdogan could again face growing domestic pressure to take unilateral military action against the PKK (see pp. 10–11).

Turkish officials will be watching closely to see how U.S. strategy toward Iraq evolves. Ankara does not want to see a precipitous withdrawal of U.S. troops from Iraq because that could lead to greater sectarian violence and draw in other outside powers—especially Iran and Syria, but possibly also Saudi Arabia. However, Turkey is adamantly opposed to increased deployment of U.S. troops in northern Iraq. Turkish officials have warned that such a move would sharply reduce Turkish cooperation with the United States and exacerbate strains in U.S.–Turkish relations.

The strains in Turkey's relations with the European Union are likely to affect U.S.–Turkish relations. In the past, when its relations with the European Union were bad, Turkey could always turn to the United States for support. But this option is no longer available. For the first time in decades, Turkey's relations with both Washington and Brussels are strained at the same time. The simultaneous deterioration of relations with the United States and the European Union has reinforced a growing sense of vulnerability and nationalism in Turkey. Turkey increasingly feels that it cannot count on the support of its traditional allies and must rely on its own devices (see pp. 22–23).

In short, the United States will need to get used to dealing with a more independent-minded and assertive Turkey—one whose interests do not always coincide with U.S. interests, especially in the Middle East. The Kurdish issue in particular could cause new divergences. How the United States handles this issue is likely to be a litmus test of the value of the U.S.–Turkish alliance in Turkish eyes. If the United States fails to take action to deal more resolutely with the PKK issue, U.S.–Turkish relations are likely to deteriorate further, and anti-Americanism in Turkey, already strong, is likely to grow.

The United States should also be careful not to present Turkey as a "model" for the Middle East, as some U.S. officials have been wont to do. This irritates many Turks, especially the Westernized elite and military, who fear that it will weaken Turkey's ties to the West and strengthen the role of Islam in Turkish politics (see p. 31). At the same time, the idea of Turkey as a model does not resonate well with the Arab states in the Middle East, which continue to resent Turkey's role as a former colonial power in the region.

The modernization of Turkish society will also pose important challenges for U.S. policy. The democratization process in Turkey over the last several decades has opened up opportunities for new groups, some of them Islamist, to enter the political arena and has eroded the ability of the traditional Kemalist elite to direct and manage Turkish foreign policy. Today, political debate in Turkey is much more open and diverse than it was 20 or 30 years ago. At the same time, as new political forces and actors enter the political arena, tensions between secularists and Islamists are likely to grow, leading to greater internal strains and political polarization (see pp. 31–21).

Acknowledgments

The author would like to thank Ian Lesser and Robert Hunter for their helpful comments on an earlier draft of this study. Any errors or misinterpretations are the sole responsibility of the author.

Abbreviations

AKP   Justice and Development Party

EU   European Union

NATO   North Atlantic Treaty Organization

PAF   Project AIR FORCE

PKK   Kurdistan Workers' Party

CHAPTER ONE

Introduction

Since joining the North Atlantic Treaty Organization (NATO) in 1952, Turkey has been an important security partner for the United States. However, Turkey's strategic importance has changed significantly in U.S. eyes since the end of the Cold War. During the Cold War, Turkey served as a barrier against the expansion of Soviet power into the Mediterranean and Middle East. Ankara tied down 24 Soviet divisions that otherwise could have been deployed against NATO forces on the Central Front. Turkey also provided important installations for monitoring and verifying Soviet compliance with arms-control agreements.

Many Turks feared that, with the end of the Cold War, Turkey would lose its strategic importance for the United States. These fears have proven unfounded. If anything, Turkey's strategic importance has increased. Turkey today stands at the nexus of three areas of critical importance to the United States: the Balkans, the Caucasus and Central Asia, and the Middle East. In each of these areas, Turkish cooperation is essential for achieving U.S. policy goals.

However, in the last decade, Turkish policy has shown a new degree of independence and activism, particularly in the Middle East. At the same time, the U.S.–Turkish security partnership has come under new strains. On a number of issues, especially policy toward Iraq, Iran, and Syria, U.S. and Turkish interests have begun to diverge. This has raised questions in some policy circles about how reliable a security partner Turkey will be in the future.

This monograph focuses on Turkey's role as a security partner for the United States. Chapter Two discusses changes in Turkey's security environment and their implications for Turkish foreign policy. Chapter Three examines key security challenges Turkey faces, while Chapter Four analyzes Turkey's most important security partnerships. Chapter Five examines the costs and benefits of the U.S.–Turkish security partnership for both sides. Chapter Six assesses the implications of all this for the United States.

CHAPTER TWO

Turkey's Changing Security Environment

The end of the Cold War had a major influence on Turkish foreign policy. During the Cold War, Turkey concentrated primarily on containing Soviet power and strengthening its ties with the West. The end of the Cold War removed the Soviet threat and opened up new opportunities and vistas to Turkish foreign policy in areas that had long been neglected or off limits to Turkish policy: the Balkans, the Caucasus and Central Asia, and the Middle East. No longer a flank state, Turkey found itself at the crossroads of a new, emerging strategic landscape that included areas where it had long-standing interests and/or historical ties. Turkey sought to exploit this new diplomatic flexibility by establishing new relationships in areas it had previously neglected, above all the Middle East and Central Asia.

In addition, the locus of threats and challenges to Turkish security has shifted. During the Cold War, the main threat came from the north—from the Soviet Union. Today, Turkey faces a much more diverse set of security threats and challenges: growing Kurdish nationalism and separatism; increasing sectarian violence in Iraq that threatens to spill over and draw in outside powers; an increasingly assertive Iran that may acquire nuclear weapons; and a weak, fragmented Lebanon dominated by radical groups with close ties with Syria and Iran. Most of these threats are on Turkey's southern periphery. As a result, Turkish attention today is focused much more intensely on the Middle East than in the past. This is where the key challenges to Turkish security are located.

At the same time, Turkey's ties with the West have deteriorated. Turkey has found its path to European Union (EU) membership blocked by rising concern in Europe about immigration, unemployment, and enlargement.[1] Cyprus has also emerged as a bone of contention in Turkey's relations with the EU. Increasingly, Turks feel unwanted and resentful at what they see as Europe's patronizing attitude toward them. As a result, Turkey's relations with the EU—and Europe generally—have become increasingly strained.

Relations with the United States have also deteriorated. The U.S. invasion of Iraq has exacerbated Turkey's security problems and strained its relations with Washington. When Turkey's relations with the EU have been strained in the past, Turkey could always look to the United States for support. Today, however, Turkey faces an unprecedented situation in which its relations with both the EU and the United States are poor *simultaneously*.

The deterioration of relations with the West has contributed to a growing sense of vulnerability and distrust of the West in parts of Turkish society. Many Turks feel that they can no longer rely on their traditional allies in the West as much as they have previously. This has reinforced a growing trend toward nationalism and a feeling that Turkey must rely more heavily on its own devices.

These trends have coincided with important domestic changes in Turkish society. The old pro-Western Kemalist elite that has shaped Turkish foreign policy since the end of World War II is gradually being replaced by a more conservative, nationalist elite that is suspicious of the West. These new elites have begun to challenge the dominance and outlook of the pro-Western elite. They are also more religious and have a more positive attitude toward Turkey's Ottoman past.

Public opinion also plays a much more important role today than in the 1960s and 1970s. As a result of the democratization of Turkish politics and society over the last several decades, new groups have

---

[1] For a comprehensive discussion, see Bulent Aliriza and Seda Ciftci, "The Train to Europe Stalls," *Turkey Update*, Washington, D.C.: Center for Strategic and International Studies, December 18, 2006. See also John Redmond, "Turkey and the European Union: Troubled Europeans or European Troubles," *International Affairs*, Vol. 83, No. 2, 2002, pp. 305–317.

entered the political arena. The foreign policy debate in Turkey is much more diverse and fluid today, with more actors influencing policy than in the past. In short, the days when the Turkish military and mandarins in the foreign ministry could control the foreign policy and security debate are over.

Finally, Turkey has witnessed a rise in religiosity over the last two decades. This has been reflected, in particular, in the growing importance of such symbolic issues as the head scarf for women. A 2006 study by the highly respected Turkish Economic and Social Studies Foundation in Istanbul, for instance, found a sharp increase in the number of respondents identifying themselves as Muslims (51 percent) rather than Turks.[2] This suggests that Turkey's Muslim identity has begun to play a more important role in the self-perception and worldview of many Turks.

This rise in religiosity has contributed to growing domestic tensions between secularists and Islamists in Turkey. These tensions were dramatized by the crisis over the election of a new Turkish president in the spring of 2007. Many secularists feared that the governing Justice and Development Party (AKP), with its strong Islamist roots, would seek to undermine the foundations of Turkey's secular order if it succeeded in winning the presidency. These concerns led to large-scale popular demonstrations by supporters of secularism in several major Turkish cities and prompted the Turkish military, the self-appointed guardians of Turkey's secular order, to issue a veiled threat of a military coup.[3]

These domestic and external trends do not mean that Turkey is about to turn its back on the West. But Turkey faces a variety of new security challenges that are pulling it more deeply into other areas, especially the Middle East, and creating new strains in relations with its Western allies, particularly the United States.

---

[2] See Ali Carkoglu and Binnaz Toprak, *Degosen Turkiye'de Din, Toplum ve Siyaset*, Istanbul: Turkish Economic and Social Studies Foundation, Yayiniari, November 2006.

[3] For a detailed discussion, see "The JDP Failure to Elect a President Triggers a New Test for Turkish Democracy," *Turkey Update*, Washington, D.C.: Center for Strategic and International Studies, May 14, 2007.

CHAPTER THREE

Security Challenges

In comparison to the Cold War period, Turkey faces a much more diverse set of security challenges. Most of these challenges are on Turkey's southern periphery. This section focuses on the most important security threats Turkey faces and how they affect Turkish foreign and security policy.

The Kurdish Challenge

The most important external challenge Turkey faces today is Kurdish nationalism. The Gulf War (1991) greatly escalated the Kurdish problem.[1] Many American policymakers view the Gulf War as the heyday of U.S.–Turkish cooperation. For many Turks, however, the war is, as Ian Lesser has noted, "the place where the trouble started."[2] The establishment of a de facto Kurdish state in Northern Iraq under Western protection gave new impetus to Kurdish nationalism and provided a logistical base for attacks on Turkish territory by Kurdish separatists in the Kurdistan Workers' Party (PKK).

---

[1] For a detailed discussion of the impact of the Gulf War on Turkish security, see F. Stephen Larrabee and Ian O. Lesser, *Turkish Foreign Policy in an Age of Uncertainty*, Santa Monica, Calif.: RAND Corporation, MR-1612-CMEPP, 2003, pp. 133–136. For background on Turkey's Kurdish problem, see Henri J. Barkley and Graham E. Fuller, *Turkey's Kurdish Question*, New York: Rowan & Littlefield Publishers, Inc., 1998.

[2] Ian O. Lesser, "Turkey, the United States, and the Delusion of Geopolitics," *Survival*, Vol. 48, No. 3, Fall 2006, p. 2.

The U.S.–led invasion of Iraq (2003) exacerbated Turkey's Kurdish problem. From the outset, Turkish leaders had strong reservations about the U.S. invasion of Iraq. They had no love for Saddam Hussein, but Saddam provided an important element of stability on Turkey's southern border. Turkish leaders feared that his removal would lead to the fragmentation of Iraq, the growth of Kurdish nationalism, and an overall decline in Turkish security.

The aftermath of the invasion has seen Turkey's worst fears come true. Iraq has degenerated into sectarian violence; Iran's influence in Iraq and regionally has increased; and the Kurdish drive for autonomy—and eventual independence—has been strengthened. As a result, Turkey today confronts the prospect that an independent Kurdish state will emerge on its southern border. Turkish officials fear this could strengthen separatist pressures among Turkey's own Kurdish population.

Since 2003, Turkey has faced an escalation of PKK-led separatist violence. The PKK has waged a guerrilla war in southeastern Turkey since 1984, often launching cross-border attacks from sanctuaries in northern Iraq. The violence subsided after the capture of PKK leader Abdullah Ocalan in 1999. But in June 2004, the PKK took up arms again. Since then, the violence has escalated dramatically. In 2006, over 600 people, many of them members of the Turkish security forces, were killed in PKK-related violence.

Prime Minister Recep Tayyip Erdogan's government has repeatedly called on the United States to provide military assistance to help eliminate the PKK threat. However, Washington has been reluctant to take military action for several reasons. First, the United States needs all available forces to fight the insurgents in Iraq and train the Iraqi security forces. Second, the United States regards the Iraqi Kurds as essential to keeping Iraq together as a unified state. If the Iraqi Kurds were to pull out of the present Iraqi coalition, the situation in Iraq might degenerate into all-out civil war. The United States has thus been reluctant to push the Iraqi Kurds too hard.[3]

---

[3] Differences within the U.S. government and U.S. military have also hindered the development of a coherent U.S. policy. The Department of State's Bureau of European and Eur-

The U.S. reluctance to take resolute action to eliminate the PKK threat—or to allow the Turks to take unilateral military measures against the PKK—has accentuated strains in bilateral relations and is one of the principal causes of the growth of anti-Americanism in Turkey. According to a German Marshall Fund poll, among Europeans, Turks have the lowest approval rating for President George W. Bush's handling of international policies, with only 7 percent approving and 81 percent disapproving. The strongest negative feelings toward U.S. leadership were also found in Turkey, where 56 percent of respondents viewed U.S. leadership as "undesirable."[4]

Turkey is also concerned about the efforts of the Kurdistan Regional Government in Northern Iraq to incorporate the city of Kirkuk and adjacent areas into areas under its control. Kirkuk sits on one of the world's largest oil deposits.[5] Several hundred thousand Kurds that Saddam Hussein had forcibly evicted as part of an effort to "Arabize" Kirkuk after the 1974 Kurdish uprising have returned to Kirkuk over the past several years to reclaim their land and homes. Turkey fears that Kurdish control of Kirkuk's oil wealth would enable the Kurds to finance an independent state. Ankara has thus opposed the Kurds' effort to "Kurdisize" the city and incorporate it into the Kurdistan autonomous region.[6] Instead, the Turks want the city to have a special status and want all ethnic groups, not just Kurds, to share power there.

Ultimately, Turkey's Kurdish problem cannot be solved through military means. It can only be resolved through a political dialogue

---

asian Affairs and the U.S. European Command have been more sympathetic to Turkish concerns, whereas the U.S. Central Command, which has military responsibility for conducting the war in Iraq, has tended to regard Turkish concerns about the PKK as a distraction.

[4] See German Marshall Fund of the United States et al., *Transatlantic Trends: Key Findings 2006*, Washington, D.C., 2006, p. 19.

[5] The known oil reserves of Kirkuk are estimated to be 12 billion barrels. However, Iraqi Kurdish officials believe an additional 10 billion barrels are there. See Cengiz Candar, "Turkey Needs to Approach Arbil for Oil Exportation," *Turkish Daily News*, March 19, 2007.

[6] For a Kurdish view, see Nouri Talabany, "Who Owns Kirkuk? The Kurdish Case," *Middle East Quarterly*, Winter 2007, pp. 1–3.

between Turkey and the Iraqi Kurdish leadership because only the Iraqi Kurdish leadership is in a position to deny the PKK assistance and sanctuary. In the 1990s, Turkey made several military incursions into northern Iraq against the PKK. None of the strikes succeeded in eliminating the PKK.

While resolving the PKK issue will not be easy, the Iraqi Kurds have a number of reasons to be interested in easing tensions with Ankara. One is economic. Northern Iraq depends heavily on Turkish trade and investment, which is estimated to be about $3 billion.[7] A decision by Turkey to curtail or stop this trade would badly damage the economy of northern Iraq.

Moreover, relations between Turkey and the Iraqi Kurds have not always been bad. During the 1990s, both Massoud Barzani, the head of the Kurdistan Regional Government in Northern Iraq, and Iraqi President Jallal Talabani closely cooperated with Turkey against the PKK.[8] Thus, enmity between Turkey and the Iraqi Kurds is by no means foreordained. Indeed, there are sound geostrategic and economic reasons for close collaboration between the two. Both sides would benefit from a reduction in current tensions.

However, while the Erdogan government favors opening a dialogue with the Iraqi Kurdish leadership, the Turkish military is opposed to opening a dialogue with the Iraqi Kurdish leaders on the grounds that the two leading Iraqi Kurdish groups, the Democratic Party of Kurdistan, headed by Massoud Barzani, and the Patriotic Union of Kurdistan, led by Iraq's President Jalal Talabani, are supporting the PKK materially and politically.[9] Given the key role the Turkish military plays in Turkish politics, especially on sensitive issues of national security, the government will need the military's support—or at least its acquiescence—for any initiative to succeed.

---

[7]   Figures provided by the Turkish Embassy, Washington D.C.

[8]   In the 1990s, Barzani and Talabani were actually issued Turkish passports, which allowed them to travel abroad under Turkish protection.

[9]   Serkan Demirtas, "Security Chiefs Nix Barzani Talks," *Turkish Daily News*, March 3–4, 2007.

However, the AKP's overwhelming victory in the July 22, 2007, elections may strengthen Erdogan's hand politically and buy him some time to pursue diplomatic initiatives aimed at reducing the PKK threat.[10] At the same time, the Turkish military knows from its experience in the 1990s that military means alone will not resolve the PKK problem. Indeed, a military strike or incursion into northern Iraq risks seriously exacerbating Turkey's difficulties. It would further strain relations with the United States and the EU and increase the number of recruits for the PKK. It could also intensify unrest among the Kurds in Turkey. The military may thus be willing to cut Erdogan some slack— at least temporarily.

However, if the PKK steps up its attacks in the aftermath of the July elections, domestic pressure could again grow for Erdogan to take military action against the PKK.

Iran

Iran presents a longer-term security challenge. Iran's growing regional influence since the U.S. invasion of Iraq is a concern in Ankara. So is the prospect that Iran might acquire nuclear weapons. At the same time, Turkey has a strong incentive to maintain good ties with Iran. The two countries share a common concern about the growth of Kurdish nationalism. This has led to an intensification of cooperation in the security field. During Erdogan's visit to Tehran in July 2004, Turkey and Iran signed a security agreement that branded the PKK a terrorist organization. Since then, the two countries have stepped up cooperation to protect their borders against guerrilla attacks by the PKK and its affiliates.

Energy is also a major driver behind the warming of Turkey's ties with Iran. Iran is the second largest supplier of natural gas to Turkey after Russia. In July 1996, shortly after taking office, Turkish Prime Minister Necmettin Erbakan concluded a $23 billion natural-gas deal

---

[10] It is noteworthy that the AKP did well in the Kurdish regions—a factor that is likely to work to the AKP's advantage in dealing with PKK issue.

with Iran. The deal set the framework for delivery of natural gas for the following 25 years. However, the deal also caused strains with the United States because it ran contrary to U.S. efforts to isolate Iran and prevent third-country investment there.

In the decade since then, energy ties have continued to expand. In July 2007, Turkey and Iran signed a memorandum of understanding to transport 30 billion m³ of Iranian and Turkmen natural gas from Iran to Europe. The deal envisages the construction of two separate pipelines to ship gas from Iranian and Turkmen gas fields. In addition, the state-owned Turkish Petroleum Corporation will be granted licenses to develop three different sections of Iran's South Pars gas field, which has estimated total recoverable reserves of 14 trillion m³.[11] These plans have drawn criticism from the United States, which continues to oppose third-country investment in Iran and favors transporting Turkmen gas by routes that avoid Iran.[12]

Turkey's growing cooperation with Iran in recent years, especially in the energy sector, highlights the degree to which U.S. and Turkish strategic perspectives in the Middle East have begun to diverge in some areas. Turkey has a strong political and economic stake in maintaining good ties with Iran. Ankara has thus been concerned about the calls for regime change in Tehran that some U.S. officials and outside specialists have made, which Ankara fears would further destabilize the Middle East. Instead, it has favored the establishment of a diplomatic dialogue with Tehran—a move advocated by the Baker-Hamilton report on Iraq.[13]

However, Turkey does not want to see the emergence of a nuclear-armed Iran. While it does not perceive an existential threat from an Iran armed with nuclear weapons, Ankara fears that Iran's acquisi-

[11] "Turkey Refuses to Back Down on Iranian Energy Deal," *Eurasian Monitor*, August 16, 2007.

[12] "U.S. Critical of Turkey's Partnership with Iran," *Turkish Daily News*, April 7, 2007.

[13] The Baker-Hamilton report was well received in Ankara in part because of its support for a postponement of the referendum on Kirkuk but also because of its emphasis on the need to engage Iran and Syria diplomatically. See James A. Baker, III, and Lee H. Hamilton, Co-Chairs, et al., *The Iraq Study Group Report*, Washington, D.C.: United States Institute of Peace, December 6, 2006.

tion of nuclear weapons could destabilize the Gulf region and force Turkey to take defensive countermeasures to safeguard its own security. Turkey has essentially three options for countering the Iranian nuclear challenge:

- expand cooperation on missile defense with the United States and Israel
- beef up its conventional capabilities, especially medium-range missiles
- develop its own nuclear capability.

The third of these would clearly be a last resort. It would only be undertaken if there were a serious deterioration of Turkey's security situation, i.e., if relations with the United States seriously deteriorated and if NATO's security guarantees no longer appeared credible. But given Turkey's current difficulties with Washington and Brussels—as well as the growing strength of nationalism in Turkey of late—the nuclear option cannot be entirely excluded.

The prospect that Iran may develop nuclear weapons is likely to heighten Turkish interest in missile defense. However, current U.S. plans to deploy elements of a missile-defense system in Poland and the Czech Republic are designed to provide protection against long-range missile threats from Iran and North Korea. They exclude Turkey and parts of southern Europe. Therefore, as it shapes its approach to missile defense in the coming decade, the United States also needs to consider how this deployment will affect Turkish security. Otherwise, current plans—which leave Turkey exposed—could exacerbate Turkish security concerns and generate new strains in U.S.–Turkish relations.

However, while Turkish officials are concerned about the long-term security implications of a nuclear-armed Iran, the Erdogan government is strongly opposed to a military strike against Tehran, which it believes could further destabilize the region. The United States could therefore not count on the use of Turkish bases in any military operation against Iran. Indeed, a U.S. military strike against Iran could provoke a serious crisis in U.S.–Turkish relations and significantly exacerbate current strains.

Syria

During much of the 1980s and 1990s, Ankara regarded Syria as a major security threat because it provided support and a safe haven for PKK terrorists. In October 1998, relations reached a crisis point when Turkey threatened to invade Syria if Damascus did not cease its support for the PKK. In the face of Turkey's overwhelming military superiority, Syria backed down and expelled PKK leader Abdullah Ocalan and closed the PKK training camps on its soil.[14]

The expulsion of Ocalan and the closing of the PKK training camps contributed to a gradual improvement in relations. This rapprochement has been driven by a mutual concern about preventing the emergence of an independent Kurdish state. (Syria also has a substantial Kurdish minority on its territory.) The intensification of ties has gained considerable momentum in the last several years, particularly since the U.S. invasion of Iraq, and Turkey has been reluctant to see these ties jeopardized.

As with Iran, Turkey's preference for engagement has conflicted with the U.S. desire to isolate Damascus and caused tensions in relations with the United States.[15] However, recent U.S. efforts to establish a dialogue with Syria may reduce frictions with Ankara and bring U.S.–Turkish approaches to Syria in closer alignment.

Greece

Turkish perceptions of Greece have shifted considerably in the last decade. As a result of the détente process initiated in 1999, relations

---

[14] For a detailed discussion of the crisis, see Yuksel Sezgin, "The October 1998 Crisis in Turkish-Syrian Relations: A Prospect Theory Approach," *Turkish Studies*, Vol. 3., No. 2, Autumn 2002, pp. 44–68.

[15] These tensions were particularly evident in spring 2005 when U.S. officials tried to pressure President Ahmet Necdet Sezer to cancel his visit to Syria. However, Sezer, supported by Erdogan, made the trip anyway, highlighting Turkey's strong stake in maintaining close ties with Damascus.

with Athens have improved significantly.[16] Consequently, today Greece is regarded as posing much less of a security challenge than it did a decade ago. However, differences over the two main security issues— the Aegean and Cyprus—remain unresolved and continue to cause periodic tensions between the two countries.

Greece and Turkey each have a strong stake in maintaining the current rapprochement. The improvement in relations has allowed both countries to focus on other challenges and contributed to greater stability in the eastern Mediterranean. However, as long as the Aegean dispute remains unresolved, there is always a danger that some incident could lead to an unwanted confrontation, as almost happened in 1996 when the two countries nearly went to war over the islet of Imia/Kardak.[17]

Cyprus also remains a source of discord. But the context and dynamics of the Cyprus issue have changed in important respects. With its admission into the EU in May 2004, Cyprus has become the major issue in Turkish–EU relations. As a result, the focus of attention on Cyprus has shifted from Washington to Brussels. This has largely removed the island as an irritant in U.S.–Turkish relations.

The Greek-Turkish détente has also reduced the saliency of the Cyprus issue as a source of friction in Greek-Turkish relations. Cyprus still stirs emotions in both countries, but the improvement in relations between Athens and Ankara has diminished the likelihood that the issue will lead to armed conflict between the two countries. This is an important shift away from the type of brinkmanship that characterized relations in the late 1990s and has contributed to greater regional stability in the Mediterranean.

---

[16] For a detailed discussion of the thaw and its motivations, see Larrabee and Lesser, 2007, pp. 84–88.

[17] In January 1996, a team of Turkish journalists removed a Greek flag from this barren islet (*Imia* in Greek, *Kardak* in Turkish) and replaced it with a Turkish flag. The incident nearly touched off a war between Greece and Turkey, which was narrowly avoided through timely U.S. mediation.

Armenia

Since the end of the Cold War, Turkey has strengthened its position in the Caucasus—a region where it has long-standing interests. Relations with Azerbaijan and Georgia have improved significantly. However, Turkey's relations with Armenia remain strained as a legacy of the massacre of Armenians by Ottoman forces in 1915–1916.

Armenia's continuing occupation of Nagorno-Karabakh poses another obstacle to better Turkish-Armenian relations. In 1993, in response to the Armenian occupation, Turkey closed its border with Armenia and suspended efforts to establish diplomatic relations with Yerevan. Turkey has made settlement of the Nagorno-Karabakh conflict a precondition for the normalization of relations with Armenia.

Recently, under U.S. pressure, Ankara and Yerevan have quietly begun to explore ways to improve relations. However, while some small progress has been made in improving relations, any major breakthrough, such as reopening the Turkish-Armenian border, is only likely after a settlement of the Nagorno-Karabakh dispute.

Islamic Extremism

Islamic extremism poses the final challenge. This challenge is largely internal but is one Turkey's Kemalist elite, especially the Turkish military, takes very seriously. The Kemalists and the military see Islamic extremism as a threat to one of the basic principles of the Turkish Republic: secularism. The election of the AKP, with its strong Islamic roots, has intensified concern among Turkey's Westernized secular elite about the Islamic challenge. Many members of the Kemalist elite fear that the AKP has a hidden Islamic agenda and that the party will eventually try to change the constitution to strengthen the role of Islam in Turkish politics and weaken Turkey's attachment to secularism.

CHAPTER FOUR

Partnerships

During the Cold War, the partnership with the United States was Turkey's most important security relationship. However, the U.S. failure to back Turkey unreservedly in the 1963–1964 Cyprus crisis prompted Ankara to reassess its foreign policy. In the wake of the crisis, Turkey began to diversify its security relationships and reduce its dependence on the United States. This process has intensified since the end of the Cold War. The U.S. partnership remains important, but it is less critical than it was during the Cold War. Because Turkey no longer faces an existential military threat from the Soviet Union, it is less in need of U.S. protection. In addition, Turkey has foreign policy options today— in the Caucasus, Central Asia, the Balkans, and the Middle East—that were not open to it several decades ago. Ankara is thus less ready to fall automatically in line behind U.S. policy, especially when U.S. policy preferences conflict with its own regional interests.

This is particularly true in the Middle East. U.S. interests and Turkish interests have increasingly diverged in recent years.[1] Iran provides an example. Here, Turkish and U.S. interests overlap only partially. Turkey and the United States share a common desire to prevent the emergence of a nuclear Iran. However, Turkey has a strong interest in maintaining good ties with Iran. Iran is a major supplier of Turkish energy, especially natural gas. Turkey and Iran also both have large Kurdish minorities and share a common interest in preventing the emergence of an independent Kurdish state. Thus, Turkey opposes

---

[1]   See F. Stephen Larrabee, "Turkey Rediscovers the Middle East," *Foreign Affairs*, Vol. 86, No. 4, July/August 2007, pp. 103–114.

U.S. efforts to isolate Iran and promote regime change in Tehran, which Ankara fears would further destabilize the region.

Similar differences exist over policy toward Syria. The United States regards Syria as a terrorist state and has sought to get Turkey to join its campaign to isolate Damascus. However, as noted earlier, Turkey needs cooperative relations with Syria to manage the Kurdish problem. It has thus opposed U.S. efforts to promote regime change in Syria, which Ankara believes would be highly destabilizing and exacerbate the Kurdish issue. Ankara has also encouraged the United States to open a dialogue with Syria and Iran, as the Iraq Study Group report advocates.

Congressional concerns about Turkey's human-rights record and Cyprus have also contributed to tensions in U.S.–Turkish bilateral relations. In recent years, the U.S. Congress has held up several important defense deals with Turkey. This has strained defense relations and contributed to the impression that the United States is a less-than-reliable defense partner. This feeling has been one of the principal driving forces behind Turkey's decision to expand defense cooperation with Israel in recent years.

Relations with the United States have also been strained by the Armenian genocide issue. In recent years, the Armenian lobby in the United States has sought to introduce into Congress a resolution that condemns Turkey for "genocide" against the Armenians in 1915–1916. In the past, successive U.S. administrations have persuaded Congress not to pass the legislation. In 2007, the Bush administration narrowly averted a serious crisis with Ankara only through intensive last-minute lobbying to prevent the genocide resolution from coming to a vote on the House floor. But the Armenian lobby, galvanized by its near success, is likely to step up its lobbying for passage of a similar bill in the future. Thus, future administrations are likely to face strong pressure to pass similar legislation.

Passage of such legislation could provoke a serious crisis in U.S.–Turkish relations. Rather than opening up dialogue between Turkey and Armenia and promoting reconciliation between the two countries, a genocide resolution would likely inflame Turkish public opinion and provoke a strong nationalist backlash. The Turkish government could come under strong domestic pressure to take retaliatory action, possibly including curtailing or halting U.S. use of Incirlik Air Base. Such action would severely hinder the ability of the United States to supply U.S. troops in Afghanistan and Iraq.

Israel

Turkey has an important strategic partnership with Israel, which has provided Turkey an alternative source of sophisticated military equipment and technology at a time of growing restrictions from the U.S. Congress on the sale of weapons to Turkey. Ankara also hoped to benefit from the Israeli lobby's influence on Capitol Hill.

However, Turkey's policy has begun to show new accents under the AKP. Erdogan has been openly critical of Israeli policy in the West Bank and Gaza, calling it an act of "state terror."[2] Turkey also hosted a high-ranking Hamas delegation in Ankara soon after the Hamas victory in the Palestinian elections. The invitation to Hamas was issued without any coordination with the United States or Israel. That strained relations with Washington and Jerusalem because it undercut the efforts of both countries to isolate Hamas until it accepted a series of conditions, including acceptance of Israel's right to exist.

The crisis in Lebanon during the summer of 2006 added to these strains. Erdogan sharply condemned Israeli attacks against Lebanon, declaring that they in no way could be considered legitimate.[3] The attacks prompted large-scale protests and the burning of the Israeli flag in several major Turkish cities. A number of Turkish nongovernmental organizations also issued statements condemning Israeli policies in Lebanon and the Palestinian territories.

The shift in Turkish policy toward Israel, however, has largely been one of tone and style. While Erodogan has been more critical of Israeli

---

[2] "Israeli Operation Draws Ire in Turkey," *The Probe*, May 23, 2004; "Turkey Irked by Gaza Offensive but Not Prompted to Reverse Ties to Israel," *The Probe*, May 30, 2004.

[3] "Erdogan: Unfair War in Lebanon Will Have No Winner," *Turkish Daily News*, August 4, 2006.

policy than his predecessors, cooperation with Israel in the defense and intelligence areas has quietly continued and been little affected by the sharper public tone in Turkey's policy. Turkey has continued to conduct trilateral military exercises with Israel and the United States.[4]

Russia

Historically, Turkey has perceived Russia as an adversary and a threat. Russia was the principal cause of the loss of Ottoman territory in the 19th century. The Cold War reinforced this adversarial relationship. However, over the last decade, cooperation between the two countries has significantly expanded, especially in the economic area.[5] Russia is today Turkey's largest trade partner and supplies more than 65 percent of Turkey's natural gas. A thriving "suitcase" trade has also developed.

The growing economic interdependence has begun to affect Turkish foreign policy and security perspectives. In the last few years, Turkey has become more sensitive to Russian security concerns and has been reluctant to adopt policies that might irritate Moscow and damage Turkish-Russian relations. This has manifested itself in particular in the cautious approach Turkey has adopted toward maritime security in the Black Sea.

Turkish and Russian perspectives on maritime security in the Black Sea closely coincide. Turkey regards itself as a Black Sea power. Like Russia, Ankara wants to preserve the military status quo in the Black Sea and opposes an increase in U.S. or NATO military presence in the region, preferring to see the Black Sea Naval Cooperation Task Group—a multilateral initiative which includes both Turkey and

---

[4] In August 2007, Turkey conducted joint search-and-rescue exercises with the United States and Israel off the coast of Turkey. See "Israel, Turkey, US to Hold Joint Military Exercises," Agence France-Presse, August 14, 2007.

[5] For a detailed discussion, see Suat Kiniklioglu, "The Anatomy of Turkish-Russian Relations," *Insight Turkey*, Vol. 8, No. 2, April–June 2006, pp. 81–96.

Russia[6]—as the main vehicle for promoting maritime security in the region.

However, there are limits to the degree of rapprochement between Turkey and Russia. The two countries are rivals for influence in Central Asia. The recent emergence of a more-assertive Russian policy in Central Asia may give this competition a sharper edge in the future. At the same time, the tougher Russian line toward cooperation with the West could make it more difficult for Ankara to improve ties with Moscow without damaging relations with its Western allies.

NATO

Unlike other southern European members of NATO, Turkey has not strongly "Europeanized" its foreign policy. This has tended to highlight Turkey's "distinctiveness" and set it apart from the rest of southern Europe.

At the same time, Turkey's geographic proximity to the Middle East colors its attitudes toward security and NATO. Turkey is the only NATO member that faces the threat of outside attack (Iran, Syria). It is thus very concerned that Article 5 (collective defense) remain a core Alliance mission and that emphasis on crisis management not weaken the Alliance's commitment to collective defense. NATO's slow response to Turkey's request for reinforcements during the 1991 Gulf War still rankles. A failure of the Alliance—or even some hesitation—to respond to a threat to Turkey could provoke a serious crisis in Ankara's ties with NATO and could lead to domestic pressures for Turkey to withdraw or suspend its membership in NATO.

---

[6] The task group is composed of naval forces from Turkey, Russia, Ukraine, Georgia, Bulgaria, and Romania. It was established in April 2001 and engages in various missions from search-and-rescue operations to environmental and humanitarian missions. For a good discussion of the security problems and interests of the key regional actors in the Black Sea region, see Eugene B. Rumer and Jeffrey Simon, *Toward a Euro-Atlantic Strategy for the Black Sea Region*, Occasional Paper 3, Washington, D.C.: Institute for National Strategic Studies, National Defense University, April 2006.

The fact that Turkey is not a member of the EU—and is not likely to become one in the near future—also colors its approach to cooperation between NATO and the EU over crisis management. While not opposed to the expansion of the EU's role in security and defense matters, Turkey does not want this to weaken NATO's role because this would reduce Turkey's ability to influence European security issues. Ankara has sought assurances that it will be involved in the planning and decisionmaking in EU crisis management operations, especially those that directly affect its own security interests.

In the last decade, differences related to Turkey's relations with the EU have increasingly spilled over into the NATO arena and affected Turkey's relations with the Alliance. Turkey held up the implementation of Berlin Plus, which allows the EU to draw on NATO assets in a crisis, for several years over fears that Greece might use its membership in the EU to push the EU to interfere in areas—particularly Cyprus—that directly affect Turkish security. Currently, differences with the EU over Cyprus (see below) are hindering the development of NATO-EU cooperation over crisis management.

Turkey's proximity to the Middle East also gives it a special interest in counterproliferation and ballistic missile defense. Turkey is the only NATO member that currently faces a threat from ballistic missiles launched from the Middle East. Iran's Shahab 3, with a range of about 1,300 km, can reach parts of eastern Turkey. Thus, as the ballistic missile threat intensifies—especially if Iran seeks to acquire nuclear weapons—Turkey is likely to show greater interest in developing a regional missile defense system with the United States and Israel.

The European Union

Membership in the EU has long been a major Turkish foreign policy goal. Turkey sees EU membership as the culmination of the centuries-long effort at Westernization and an affirmation of the Kemalist revolution. However, the EU remains ambivalent about Turkish membership. At its summit in Brussels in December 2004, the EU agreed to open accession negotiations with Turkey. However, the EU combined its assent with a number of caveats and escape clauses, noting that the initiation of negotiations did not guarantee their successful completion or preclude other forms of association short of membership.

Turkey's membership prospects have dimmed since the Brussels decision. The rejection of the Constitutional Treaty by France and the Netherlands in May and June 2005 made clear that large parts of the European public think the EU is moving too far, too fast. Bulgaria and Romania were admitted in January 2007. However, further enlargement is likely to be put on hold for some years, as EU members seek to develop a consensus about the EU's future evolution and priorities.

The Cyprus issue also complicates Turkey's relations with the EU. Under the Customs Union agreement signed with the EU in 1996, Turkey is obligated to open its ports and airports to Cypriot vessels and aircraft now that Cyprus is a full member of the EU. However, Turkey has refused to do so until the EU fulfills its promise to lift its trade embargo against Northern Cyprus. In response, in December 2006, the EU Council voted to suspend eight out of 35 chapters in the accession negotiations. As a result, relations with the EU have become strained.

At the same time, frustration with and anger toward the EU is rising in Turkey. Support for Turkish membership in the EU has declined visibly over the last year. In 2004, 73 percent of the Turkish population supported Turkish membership; in 2006, that portion dropped to 54 percent.[7] This decline reflects a significant erosion of support for Turkish membership in the EU and illustrates how the public mood in Turkey toward the EU has soured of late.

In the past, when relations with the EU were bad, Turkey could always turn to the United States. But this option is no longer as attractive. As noted at the outset, for the first time in decades, Turkey's relations with both Washington and Brussels are strained at the same time. The simultaneous deterioration of relations with the United States and the EU has reinforced a growing sense of vulnerability and nationalism in Turkey. Ankara increasingly feels that it cannot count on the support of its traditional allies and must rely on its own devices.

---

[7] German Marshall Fund of the United States, 2006, p. 19.

The deterioration of relations with the United States and Europe has not only reinforced a growing sense of nationalism but also sparked a search for new options. Some parts of the AKP have begun to look to the Middle East as a means of compensating for weakening relations with the West, while others have suggested that Turkey should look more to Russia. Some have also advocated a "Eurasian option," which would exploit Turkey's economic and cultural ties with the former Soviet republics of Central Asia.

This is not to suggest that Turkey is likely to turn its back on the West or abandon its quest for EU membership. Economically, Turkey remains closely tied to the West, especially Europe. (More than half its overall trade is with Europe.) But it does illustrate the degree to which Turkey's relations with the West are under stress and the way in which Turkey is beginning to diversify its policy.

CHAPTER FIVE

Benefits and Costs

Washington and Ankara continue to see important benefits in maintaining a close security partnership. However, new strains have emerged—especially since September 11, 2001—that have eroded the strength and robustness of the partnership.

U.S. Perspectives

The United States regards Turkey as an important strategic ally. However, the context has changed. During the Cold War, Turkey was important as a barrier against the expansion of Soviet power into the Middle East. Today, Turkey's strategic importance lies in its capacity to act as a bridge to the Muslim world and serve as a stabilizing force in the Middle East and Central Asia—two areas of growing strategic importance to the United States.

Turkey's strategic importance was underscored during the 1991 Gulf War. Ankara granted access and overflight rights to U.S. combat and aircraft operating from Incirlik Air Base and elsewhere in Turkey. Ankara deployed some 100,000 troops along the Iraqi border, pinning down substantial Iraqi forces. Turkey also shut down its pipelines, cutting off Iraqi oil exports. And after the conclusion of the war, it allowed allied aircraft to fly sorties out of Incirlik to monitor the no-fly zone over northern Iraq.

However, in recent years, the Turks have increasingly restricted the use of Incirlik. They have allowed the United States to use Incirlik to resupply troops in Afghanistan and Iraq but are highly sensitive about the use of the base for combat missions in the Middle East. The United States is thus unlikely to be able to use Turkish bases for operations other than those spelled out in 1980 Defense and Economic Cooperation Agreement.

The growing divergence between U.S. and Turkish perspectives in the Middle East has reinforced Turkish sensitivity about granting the United States access to Turkish territory—as the Turkish parliament's refusal to allow the United States to use Turkish territory to open a second front against Iraq in March 2003 underscores. While the issue was poorly managed on both sides, the refusal reflects deep-seated Turkish concerns about national sovereignty that are increasingly likely to influence Turkish attitudes toward the use of its bases for future contingencies in the Middle East.

The United States has strong strategic interest in ensuring that this cooperation continues. Currently, over 70 percent of military personnel and material destined for Iraq passes through Turkish territory. If Turkey were to curtail U.S. use of Incirlik for any reason, the effect on the ability of the United States to conduct operations against the insurgents in Iraq and support the Iraqi government would be significant. Moreover, many of the withdrawal scenarios currently under discussion would require Turkish cooperation.

Turkish Perspectives

For Turkey, the security relationship with the United States also remains important. Turkey lives in a tough and volatile neighborhood and has disputes with several neighbors (Iraq, Greece, Armenia). It also is within range of missiles fired from Iran and Iraq. Turkey thus views its security relationship with the United States as an important insurance policy against its growing exposure to risks in the Middle East. While U.S. involvement in the Middle East also entails risks for Turkey, Turkey benefits on balance from the U.S. alliance and its military presence in adjacent regions.

The United States is also Turkey's most important arms supplier. Despite recent efforts at diversification, Turkey still conducts roughly

80 percent of its defense-industrial activity with the United States. Large numbers of Turkish officers have been trained in the United States. This has allowed the Turkish military personnel to develop close ties with their American counterparts and develop a deeper knowledge of American military operational doctrine and thinking.

Finally, the United States has strongly supported key Turkish strategic priorities outside the defense realm. For example, construction of the Baku-Ceyhan oil pipeline is a key Turkish strategic priority designed to bring Caspian oil to world markets via a terminal on Turkey's Mediterranean coast. Washington has also strongly backed Turkey's bid for EU membership and supported Turkey's struggle against the PKK separatists much more vigorously than have Ankara's European allies.

These factors have underscored the benefits of maintaining close security links to the United States. However, Turkish perceptions regarding the value of the U.S.–Turkish security partnership have shifted markedly in the last several decades.

During the Cold War, the feeling that Turkey derived important benefits from its security relationship with the United States was widespread among the Turkish elite and population alike. This perception began to change after the Cyprus crisis in 1963–1964. The famous "Johnson Letter"—in which President Lyndon Johnson warned that the United States might not come to Turkey's defense if Turkish intervention in Cyprus provoked a Soviet response—came as a shock to the Turks. The crisis underscored that there were costs to being so heavily dependent on the United States and prompted Turkey's effort to broaden its security ties and reduce its dependence on Washington.

The perception that maintaining close security ties with Washington had important costs was reinforced in 1975 when the United States imposed an arms embargo on Turkey in response to the Turkish invasion of Cyprus. Turks regarded the embargo as a slap in the face of a loyal ally, so the embargo led to a sharp deterioration of U.S.–Turkish relations. It is still remembered with bitterness and colors contemporary Turkish attitudes about the degree to which the United States can be considered a reliable ally.

The 1991 Gulf War marked an important watershed in Turkish perceptions of the costs of unswerving support for the United States. Contrary to the advice of most of his top civilian and military advisors, President Turgut Özal threw his full support behind the U.S. campaign to expel Saddam Hussein from Kuwait. According to Turkish estimates, Turkish support in enforcing sanctions against Iraq cost Turkey $6 billion in lost revenue, and the deployment of 100,000 troops along the Iraqi border cost Turkey another $300 million.

Özal expected Turkey's support to strengthen its "strategic partnership" with the United States significantly and enhance Turkey's chances of obtaining membership in the EU. Neither of these expectations was fulfilled. The financial losses Turkey incurred and the lack of tangible benefits from its support for the United States in the Gulf War contributed to a growing perception in Ankara that Turkey gets much less from the relationship than the United States does.

This perception has been reinforced by the reluctance of the United States to take military action to help Turkey eliminate the threat the PKK poses. For many Turks, the PKK is the litmus test of the value of the U.S.–Turkish security partnership. If the United States fails to address Turkish concerns about the PKK more resolutely, strains in U.S.–Turkish relations are likely to increase, and security in the Middle East will become even more precarious.

CHAPTER SIX

Implications for the United States

In the future, Turkey is likely to be a less predictable and more difficult ally. While it will continue to want good ties with the United States, Turkey is likely to be drawn more heavily into the Middle East by the Kurdish issue, Iran's nuclear ambitions, and the fallout from the crisis in Lebanon. As a result, the tension between Turkey's Western identity and its Middle Eastern orientation is likely to grow. At the same time, the divergences between U.S. and Turkish interests that have manifested themselves over the last decade are likely to increase.

Given its growing equities in the Middle East, Turkey is likely to be even more reluctant in the future to allow its bases, particularly Incirlik, to be used to undertake combat operations in the Middle East. President Özal's willingness to allow the United States to fly sorties out of Incirlik during the Gulf War was the exception, not the rule. Since then, Turkey has increasingly restricted U.S. use of Incirlik for combat missions in the Middle East. The United States should therefore not count on being able to use Turkish bases, particularly Incirlik, as a staging area for combat operations in the Gulf and Middle East.

Moreover, given the importance of the Kurdish issue for Turkish security, Turkey has strong reasons to pursue good ties with Iran and Syria, both of which share Turkey's desire to prevent the emergence of an independent Kurdish state. Turkey's growing energy ties with Iran have reinforced interest in that particular tie. Thus, Turkey is unlikely to support U.S. policies aimed at isolating Iran and Syria or overthrowing the regimes in either country. Rather, Ankara is likely to favor policies aimed at engaging Iran and Syria and to encourage the United States to open dialogues with both countries.

At the same time, Turkey does not want to see a nuclear-armed Iran, which Ankara fears could spark a nuclear arms race in the Gulf region and the Middle East. Here, U.S. and Turkish interests overlap. Turkish concerns about Iran's possible desire to acquire nuclear weapons are likely to intensify if negotiations aimed at obtaining Iranian compliance with the International Atomic Energy Agency's regulations stall or collapse. This could open up new opportunities for U.S.–Turkish cooperation, especially in missile defense.

In the near term, however, the most important potential source of discord between the United States and Turkey is likely to be over how to deal with the terrorist attacks the PKK conducts from sanctuaries in northern Iraq. The number of Turkish security forces that the PKK has killed has risen dramatically in the last several years. Domestic pressure, especially from the Turkish military, has been growing for Turkey to take unilateral military action against the PKK. The AKP's landslide victory in the July 22, 2007, parliamentary elections has strengthened Erdogan's hand politically and bought him some breathing room diplomatically. But if the PKK attacks intensify in the aftermath of the elections, Erdogan could again face growing domestic pressure to take unilateral military action against the PKK.

Turkish officials will be watching closely to see how U.S. strategy toward Iraq evolves. Ankara does not want to see a precipitous withdrawal of U.S. troops from Iraq because that could lead to greater sectarian violence and draw in other outside powers—especially Iran and Syria but possibly also Saudi Arabia. However, Turkey is adamantly opposed to increased deployment of U.S. troops in northern Iraq, as some former U.S. officials have advocated.[1] Turkish officials have warned that such a move would sharply reduce Turkish cooperation with the United States and exacerbate strains in U.S.–Turkish relations.

---

[1] See Ronald D. Asmus and Richard C. Holbrooke, "Re-reinventing NATO," presented at 2006 Riga Conference, Riga, Latvia, 27–29 November 2006, Washington, D.C.: German Marshall Fund of the United States, 2006, p. 5.

In short, the United States will need to get used to dealing with a more independently minded and assertive Turkey—one whose interests do not always coincide with those of the United States, especially in the Middle East. The Kurdish issue in particular could cause new divergences. How the United States handles this issue is likely to be a litmus test of the value of the U.S.–Turkish alliance in Turkish eyes. If the United States does not address Turkish concerns about the PKK more resolutely, U.S.–Turkish relations are likely to further deteriorate, and anti-Americanism, already strong, is likely to grow.

The United States should also avoid portraying Turkey as a "model" for the Muslim countries in the Middle East, as some U.S. officials have been wont to do. This irritates many Turks, especially the Westernized elite and military, who fear that it will weaken Turkey's ties with the West and strengthen the role of Islam in Turkish politics. At the same time, the idea of Turkey as a model does not resonate well with the Arab states in the Middle East, which continue to resent Turkey's role as a former colonial power in the region.

The modernization of Turkish society will also pose important challenges for U.S. policy. The democratization process in Turkey over the last several decades has opened up opportunities for new groups, some of them Islamist, to enter the political arena and has eroded the ability of the traditional Kemalist elite to direct and manage Turkish foreign policy. Today, political debate in Turkey is a much more open and diverse than it was 20 or 30 years ago. At the same time, as new political forces and actors enter the political arena, tensions between secularists and Islamists are likely to grow, leading to greater internal strains and political polarization.

Turkey also faces an important change of political leadership. The older political leaders who guided Turkish policy for much of the Cold War era, such as Sulyman Demirel and Bulent Ecevit, are fading from the political scene. Policymakers in the United States will need to reach out to a new generation of Turkish politicians whose worldview will be quite different both from that of the United States and that of their predecessors.

Finally, the role of the Turkish military is changing. For the past 70 years, the military has acted as the "custodian of Turkish democracy," exerting a strong political role behind the scenes and intervening when it thought democracy and secularism were threatened. The democratic reforms introduced in the past decade or so, especially those the Erdogan government has introduced, have reduced the power of the military to intrude in politics.[2] However, as the military's "midnight memorandum" at the end of April 2007 underscores, the military continues to regard itself as the ultimate guardian of Turkey's secular order.[3] Whether the military will be willing to accept a significantly diminished political role, as required for EU membership, is an open question and likely to be one of the critical issues affecting Turkey's political evolution in the coming decade.

---

[2] For a detailed discussion, see Gareth Jenkins, "Continuity and Change: Prospects for Civil-Military Relations in Turkey," *International Affairs*, Vol. 83, No. 2, 2007, pp. 339–355.

[3] The veiled threat of a coup in the April 27, 2007, memorandum from the Turkish General Staff might seem at first glance to suggest that the military might play a larger political role in the future. However, the negative public reaction to the memorandum—reflected in the press and the slogan of many street demonstrations, "neither Sharia nor coup"—caught the military leaders by surprise and is likely to make the them cautious about intervening in the political process in the future.

Bibliography

Aliriza, Bulent, and Seda Ciftci, "The Train to Europe Stalls," *Turkey Update*, Washington, D.C.: Center for Strategic and International Studies, December 18, 2006.

Asmus, Ronald D., and Richard C. Holbrooke, "Re-Reinventing NATO," presented at the 2006 Riga Conference, Riga, Latvia, 27–29 November 2006, Washington, D.C.: German Marshall Fund of the United States, 2006.

Baker, James A., III, Lee H. Hamilton, et al., *The Iraq Study Group Report*, Washington, D.C.: United States Institute of Peace, December 6, 2006.

Barkley, Henri J., and Graham E. Fuller, *Turkey's Kurdish Question*, New York: Rowman & Littlefield Publishers, Inc., 1998.

Carkoglu, Ali, and Binnaz Toprak, *Degosen Turkiye'de Din, Toplum ve Siyaset*, Istanbul: TESEV Yayiniari, November 2006.

Candar, Cengiz, "Turkey Needs to Approach Arbil for Oil Exportation," *Turkish Daily News*, March 19, 2007.

Demirtas, Serkan, "Security Chiefs Nix Barzani Talks," *Turkish Daily News*, March 3–4, 2007.

"Erdogan: Unfair War in Lebanon Will Have No Winner," *Turkish Daily News*, August 4, 2006.

German Marshall Fund of the United States, et al., *Transatlantic Trends: Key Findings 2006*, Washington D.C., 2006.

"Israel, Turkey, US to Hold Joint Military Exercises," Agence France-Presse, August 14, 2007.

"Israeli Operation Draws Ire in Turkey," *The Probe*, May 23, 2004.

"The JDP Failure to Elect a President Triggers a New Test for Turkish Democracy," *Turkey Update*, Washington, D.C.: Center for Strategic and International Studies, May 14, 2007.

Jenkins, Gareth, "Continuity and Change: Prospects for Civil-Military Relations in Turkey," *International Affairs*, Vol. 83, No. 2, 2007.

Kiniklioglu, Suat, "The Anatomy of Turkish-Russian Relations," *Insight Turkey*, Vol. 8, No. 2, April–June 2006, pp. 81–96.

Larrabee, F. Stephen, "Turkey Rediscovers the Middle East," *Foreign Affairs*, Vol. 86, No. 4, July/August 2007, pp. 103–114.

Larrabee, F. Stephen, and Suat Kiniklioglu, "Outside View: America's Turkey Problem," United Press International, February 23, 2007.

Larrabee, F. Stephen, and Ian O. Lesser, *Turkish Foreign Policy in an Age of Uncertainty*, Santa Monica, Calif.: RAND Corporation, MR-1612-CMEPP, 2003. As of October 3, 2007:
http://www.rand.org/pubs/monograph_reports/MR1612/

Lesser, Ian O., "Turkey, the United States, and the Delusion of Geopolitics," *Survival*, Vol. 48, No. 3, Fall 2006, p. 2.

Lesser, Ian O., *Beyond Suspicion: Rethinking U.S.-Turkish Relations*, Washington D.C.: Woodrow Wilson Center for Scholars, 2007.

Redmond, John, "Turkey and the European Union: Troubled Europeans or European Troubles," *International Affairs*, Vol. 83, No. 2, 2002.

Rumer, Eugene B., and Jeffrey Simon, *Toward a Euro-Atlantic Strategy for the Black Sea Region*, Occasional Paper 3, Washington, D.C.: Institute for National Strategic Studies, National Defense University, April 2006.

Sezgin, Yuksel, "The October 1998 Crisis in Turkish-Syrian Relations: A Prospect Theory Approach," *Turkish Studies*, Vol. 3., No. 2, Autumn 2002.

Talabany, Nouri, "Who Owns Kirkuk? The Kurdish Case," *Middle East Quarterly*, Vol. XIV, No. 1, Winter 2007.

"Turkey Irked by Gaza Offensive but Not Prompted to Reverse Ties to Israel," *The Probe*, May 30, 2004.

"Turkey Refuses to Back Down on Iranian Energy Deal," *Eurasian Monitor*, August 16, 2007.

"U.S. Critical of Turkey's Partnership with Iran," *Turkish Daily News*, April 7, 2007.